The Electron Accelerator Laboratory

at

Yale University

1946 – 1986

Frank W. K. Firk

Professor Emeritus of Physics. Yale University

2018

ISBN − 13:978-1721777662

ISBN − 10:1721777660

Printed in USA by CreateSpace, Charleston, SC

Contents

Note that equation numbers (1) . . . (N) begin anew in each Chapter.

Preface

It is all too easy to forget the immediate past; this is particularly true in research fields that involve rapidly developing technologies. Physics is one such field. A case can be made that the fields of classical Atomic, Nuclear and Particle Physics have all passed their heydays; they have been remarkably successful – our knowledge of these subjects is deep and such that the remaining outstanding problems in them are relatively few in number. It is important, however, not to forget the great strides that were made in our understanding of the fundamental issues associated with the three fields. It is with these thoughts in mind that the present historical survey has been written. The material covers an important period in the history of Physics at Yale – a period in which so much was there to be discovered, and so much was there to be done in a do-it-yourself fashion.

In this brief book, the subject matter has been limited to studies of neutron and photon interactions with nuclei Electron scattering studies have not been covered; it would have required another book to do the field justice.

1. Introduction

In 1924, G. Ising published a paper entitled "The Principle of a Method for the Production of Canal Rays (a beam of positive ions) of High Voltage". His proposal involved passing the beam into a sequence of drift tubes with gaps between them; an electrical wave-front generated by a spark discharge, appeared across successive gaps in a delayed fashion, so that the electrical field occurred across a given gap at the instant the positive ions arrived, thus causing the ion beam to be accelerated. For practical reasons, he never constructed such a device.

It was R. Wideroe (1928) who designed, built and operated the first ion accelerator. Wideroe used a series of drift tubes that were connected to a source of high frequency electromagnetic waves. The drift tube lengths, and the frequency, were chosen so that the ions arrived at successive gaps when the fields were in an accelerating phase. In the 1930's there were no serious developments of linear accelerators because there were no sufficiently powerful sources of electromagnetic radiation.

In 1945, Professors E. Robert Beringer and Howard L. Schultz returned to Yale University after playing key roles in the development of Radar during World-War II. They were experts in electronics, particularly high-power, microwave systems. They immediately began doing research in the field of particle accelerators that were necessary in the study of nuclear structure – at the time, the most fashionable area of Physics. Professor Beringer began work on the design and construction of a heavy ion accelerator, and Professor Schultz began work on the design and construction of a linear electron accelerator. As a result of their original and imaginative research, they built particle accelerators that played major roles in research in Nuclear Physics carried out in the Department of Physics from the mid-1940's to the mid-1980's. In this brief history, key aspects of the wide-ranging research carried out at the Electron Linear Accelerator Laboratory (1946-1986) are discussed.

Professor Schultz's original accelerator is shown in figure 1.

Fig.1 Professor Schultz's original cavity linear electron accelerator. Construction took place in the Physics Department at Yale, beginning in 1946.

The 10 to 70 MeV traveling wave linear electron accelerator officially opened in 1962, see figures 2 and 3.

DEDICATION OF THE

LINEAR ELECTRON ACCELERATOR

Saturday, February 17, 1962

YALE UNIVERSITY, NEW HAVEN, CONNECTICUT

Fig. 2 Official dedication of the Yale Electron
Linear Accelerator, February, 1962

10

PROGRAM

PRESIDING

Howard L. Schultz, *Professor of Physics and Director of the Electron Accelerator Laboratory*

PRESENTATION

Leland Haworth, *Member of the Atomic Energy Commission*

ACCEPTANCE

A. Whitney Griswold, *President of Yale University*

Fig. 3 Acceptance of the Electron Linear Accelerator on behalf of the University by the President, A. Whitney Griswold.

This was an important occasion in the history of the Department. A view of the accelerator, from the target end, is shown in figure 4.

11

Fig. 4 The 10 to 70 MeV traveling-wave, electron linear accelerator in its massive concrete housing (1962); it was constructed by the High Voltage Engineering Corporation of Walnut Creek, CA.

2. Electron Linear Accelerators

2.1 Principles of Operation

The following outline of the fundamental principles of traveling wave linear accelerators is given; they involve applications of Maxwell's Theory of Electromagnetism not usually covered in Graduate courses.

A circular waveguide is required that propagates a wave with an axial component of electric field. The intention is to introduce a monoenergetic beam of electrons at the input of the guide and to form bunches of electrons that surf-ride in an accelerating field. The electrons are to be maintained in phase with the wave by increasing the phase velocity of the wave at the same rate as the average electron velocity.

Ordinary waveguides cannot be used in this way because they have a phase velocity that is always greater than c, the velocity of light in free space. However, a waveguide with deep corrugations in the outer wall can have a phase velocity less than c; magnetic energy is stored in the structure. The decrease in phase velocity is analogous to the inductive loading of a transmission line in a delay network.

The theory must relate the phase velocity of the wave to the radio frequency and the geometrical parameters of the corrugated guide. The simplest EM wave with a phase velocity < c, and an axial electric field is a circularly symmetric solution of the EM field equations with axial and radial components of the electric field, and a circular magnetic field. The relevant equations are then (see Stratton, 1941)

$$E\rho = E_0(i\beta/\chi)J_1(\chi\rho)\Phi$$
$$E_z = E_0 J_0(\chi\rho)\Phi \qquad \ldots\ldots\ldots (1)$$

and

$$Z_0H_\phi = E_0(ik/\chi)J_1(\chi\rho)\Phi$$

where ρ, ϕ, z are cylindrical coordinates, Z_0 is the intrinsic impedance of free space, k is the wave number (2π/free space wavelength), χ and β are the components of the phase constant in the ρ and z directions, $k^2 = \chi^2 + \beta^2$ and $\Phi = \exp\{-i\beta z + i\omega t\}$. The J's are Bessel functions.

For a phase velocity $< c$ we have $\beta > k$, χ is now imaginary and E_z increases with radial distance.

Cutler (1944) was the first person to develop the theory in a useful way. He assumed that the field in the axial region is given by equations (1) and that the radial field impedance is of the form

$$|E_\rho/H_\phi| = Z_0(i\chi/k)\{J_0(\chi\rho)/J_1(\chi\rho)\}. \quad (2)$$

He equated this impedance to the impedance looking into the mouths of the corrugations, and he assumed that the tangential field across the mouth of the nth corrugation at $\rho = a$ to be

$$E'_z = \{Cexp(-i\beta_0 nD)\}/\{1 - ((2x_n/d)^2\}^{1/2}, \quad (3)$$

and zero over the intervening metal walls. Here, a = radius of inner boundary of guide, D = corrugation pitch, d = width of corrugation mouth, $\beta_0 = 2\pi/\lambda_g$ where λ_g is the guide wavelength, and x_n is measured from the center of the nth corrugation.

The phase change per corrugation width is $\beta_0 D$, therefore, β can only have values given by

$$\beta_m D = \beta_0 D + 2\pi m. \quad (4)$$

For $\rho < a$, the component E_z is then given by

$$E_z = \Sigma_{-\infty,\infty} A_m J_0(\chi_m \rho)exp[-i\beta_m z + i\omega t] \quad (5)$$

where $A_m = (C\pi d/2)J_1(\beta_m d/2)/J_0(\chi_m a)$. (6)

Placing these values of A_m in the field equations (including $E\rho$ and $E\phi$) gives the complete field equations in the region $\rho < a$.

Over the decades, Cutler's approach has been refined, but it remains, to this day, the fundamental theory of traveling wave accelerators.

A similar analysis is used to obtain the fields inside the corrugations. To a sufficient degree of accuracy, Cutler found

$$E_z = BF_0(k\rho)G \tag{7}$$

and

$$Z_0H\phi = iBF_1(k\rho)G \tag{8}$$

where $B = C\pi/2F_0(ka)$ and $G = \exp\{-i\beta_0 nD\}$. The values of F_0 and F_1 are given in terms of Bessel functions. The ratio $F_1(ka)/F_0(ka)$ results in a *frequency equation* that gives the relationship among guide dimensions, wave velocity and frequency.

In Cutler's approximation, the fields inside the corrugations are transmission line solutions with a constant phase shift in the fields between adjacent sections.

2.2 The Yale Electron Accelerators

2.2.1 Schultz's accelerator

Beginning in 1946, Professor Schultz designed, built and used a linear electron accelerator of a unique kind. The radiofrequency power was provided by triodes with peak powers of 500 kW, operating at frequencies of 600 MHz. The design and construction of the accelerator was carried out in the Department at Yale. In its final form (8 cavities) the electron energy achieved was 7MeV. The 8 RF cavities were synchronized to accelerate electrons from their injection energy of 50keV. Each pulse was typically 1μsec in duration. Pulses were produced at a rate of 90 pulses per second.

A photo of the 8-cavity accelerator is shown:

Fig. 5 The 8-cavity linear electron accelerator designed and built at Yale in 1950.

The pulses of electrons were used to produce bursts of neutrons via the Be(γ, n) reaction. The production rate of neutrons in a pulse was 10^{14} neutrons per sec. The energy spectrum of the neutrons was Maxwellian that peaked at 1MeV. The spectrum was moderated, and the interactions of low-energy neutrons with nuclei were studied using the time-of-flight method with a 10-meter flight path.

The major part of the research involved studies of the spectra of gamma rays that followed neutron capture in heavy elements. Of particular note were those experiments in which coincidences between gamma rays in the decaying spectra were detected. These measurements were the first of their kind, and provided information on the spins and parities of the states involved.

In 1951, a proposal was made to the Atomic Energy Commission asking for funds to build and use a new 20MeV linear electron accelerator at Yale that would exceed the performance of a newly installed traveling wave accelerator at the Atomic Energy

Research Establishment at Harwell in England. (At the time, Professor Firk was working on the machine in England; it was not until 1965 that he came to Yale to continue his research in the fields of neutron and photoneutron physics). The Yale proposal was not approved because it was clear that future models of electron linear accelerators would be of the traveling-wave variety. In the next section, a powerful machine is described; it was installed six years after negotiations began between Yale and the Government in 1955. (It should be noted that the 1951 proposal was made by Professor Schultz and Professor Gregory Breit — how many renowned theoretical physicists have been so intimately involved in such a proposal?).

2.2.2 The 10 –70MeV accelerator

In the mid-fifties, it became clear that, in order to be in the forefront of research in

Nuclear Physics a new, more powerful and higher-energy electron accelerator would be required. At that time, Professor Charles Bockelman arrived and immediately joined Professor Schultz in writing a proposal to the AEC for an accelerator with energies in the range 10 to 70 MeV. It was designed for use in studies of neutron, photon, and electron interactions with nuclei. The cost of the accelerator was about 1 million dollars, and a new large laboratory was required; the laboratory was to be funded by the University. A new electron accelerator was installed by 1961.

The essential features of the world-class accelerator were:

Number of klystron modulators: 5

Klystron frequency: 1.3 GHz

Peak power of a modulator :   5 MW

Maximun operating frequency: 500 pps

Electron injection energy: 100 keV

Electron pulse durations: 5nsec to 5μsec

Electron output energy: 10 to 70 MeV

Intrinsic electron energy spread ≈ 5%

Best analyzed energy spread: 0.02%

In 1969, the accelerator laboratory had a non-academic staff of 13 persons:

1 chief engineer (Phil Jewett, later Bob Carr)

1 RF specialist (Walt Knudsen)

3 accelerator operators (Frank Hegedus, Sidney Clow, Marv Garfield)

3 machinists and laboratory technicians (Ed Comeau, Al Comeau, Joe Cimino)

3 draftsmen and designers (Joe Berti, Al Nelson, Kathy Lappert)

1 business manager (Russ Jones)

1 secretary (Bernadette Grieb . . .)

The accelerator operated 24 hours per day, 5 days per week for 11 months each year. The downtime was remarkably low for such a complex machine.

In 1969, the annual budget, funded by a Government Agency, was $550,000.

Details of Faculty and Research members of the Laboratory are given in Chapter 5. The following photo (figure 6), taken by a reporter, shows Professors Bockelman, Draper and Schultz in the control room of the new accelerator:

MONITORING GIANT ATOM SMASHER—Closed circuit television helps monitor the controls of Yale's huge new linear accelerator. Three of the Yale scientists who will direct research are shown. They are (left to right) Charles K. Bockelman, James E. Draper and Howard L. Schulz, director. The accelerator rests some 75 feet away from the controls.

The "modulator room" with 5 high-powered klystrons is shown in the following photo:

Fig. 7 The klystron modulator room. The five vertical, high-powered RF klystron modulators were mounted above their oil-filled, 250kV pulse transformers. The thyratron modulator drive circuits were housed in the metal-shielded cages. Each klystron assembly produced a peak power of 5 MW in a 5μsec pulse. At full power, the system operated at 500 pulses per second, with a beam power, on target, of 35kW. The (black) waveguides carried the RF power to the five sections of the traveling wave accelerator. The klystrons operated at 1.3 GHz (L-band).

2.3 Technical developments

2.3.1 Nanosecond electron gun injector

The original accelerator, manufactured by the High Voltage Engineering Corporation, had a minimum electron pulse width of 1μsec. To carry out high-resolution time-of-flight experiments using neutrons with energies in the MeV-region it was necessary to reduce the pulse width to a few nanoseconds.

The circuit, shown in figure 8,   produced electron beams with widths ≈ 5 nsec. It is a triggered spark gap. The high-voltage, high-current thyraton, generated a pulse with an amplitude > 10kV. The spark gap, with a short-circuited delay line, produced a pulse duration of 5 nsec. A weak radioactive source close to the gap was found to stabilize the time-jitter in the pulse.

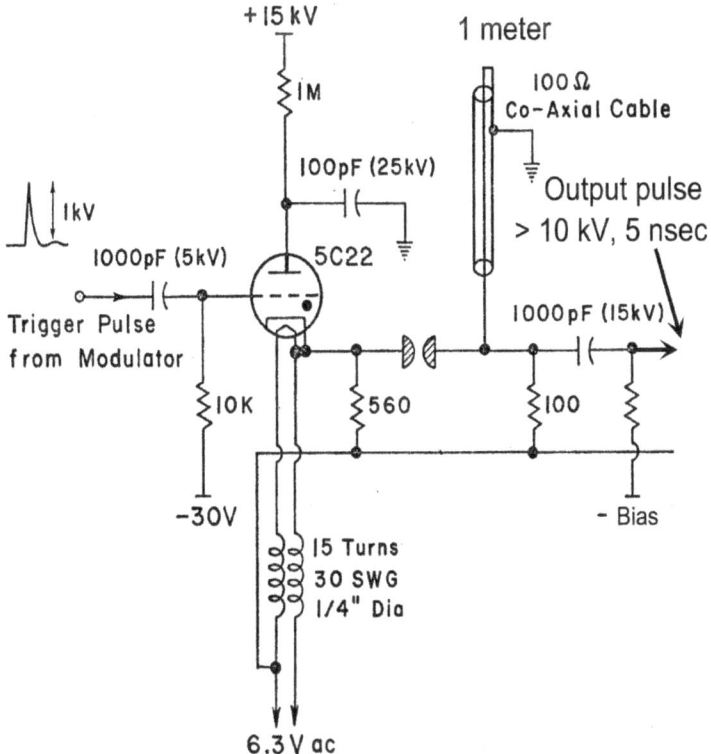

Fig. 8 The nanosecond electron gun modulator. When driving a high-current, triode electron gun, pulses of 100keV electrons were injected into the accelerator with peak currents >10A and with durations ≈ 5nsec.

The original injector consisted of a pair of deflector plates that chopped a continuous beam of 100 keV electrons into short pulses; the system was replaced with a high-current triode electron gun designed with coaxial characteristics. The modulator shown in figure 8 generated pulses 5nsec in duration with peak currents > 10 A. The system ran reliably for more than a decade.

2.3.2 Double-pulsed injector system

The original accelerator was delivered with an electron injection system that operated at a potential difference of 100kV with respect to ground potential.

This required an insulated, pressurized gas-filled tank to house all the injection electronics and control apparatus. The gases resulted in corrosion of internal materials, and therefore, in unnecessary accelerator downtime. In the early 1970's, Professors Schultz and Firk designed and built a unique double-pulsed injector system in which the triode gun was at ground potential except for a period of 1.5 μsec (the filling time of the waveguides in the accelerator) when a

-100kV pulse was applied to the gun. During the 1.5 μsec interval, a 5nsec pulse was applied to the grid of the triode. In this way, the entire injector system operated in the normal atmospheric conditions of the Laboratory. The system worked flawlessly for the next decade.

3. Neutron Physics

3.1 Introduction

The first studies of low-energy neutron induced nuclear reactions in heavy nuclei, carried out in the mid-1930's, showed narrow resonant states at excitation energies of more than 5 MeV in a typical nucleus. The observation of these states led Niels Bohr (1936) to introduce the compound nucleus model in which an incident neutron enters the potential well of the target nucleus and is trapped. The compound state does not decay until sufficient energy resides in one of the nucleons for it to escape, or for it to decay by emitting a sequence of gamma rays. This is a statistical process and therefore it can take a long time (on a nuclear time scale) for the compound system to give up its excitation energy.

(A narrow state with a energy width of 1 eV has a lifetime of 10^{-15} seconds).

Breit and Wigner (1936) introduced a theory of the resonant form. Kapur and Peierls (1937) and Wigner and Eisenbud (1947) generalized the theory. The Wigner-Eisenbud version has been used to analyze almost all nuclear reaction data since that time.

In the period between 1955 and 1960 the observed level spacing distribution of neutron resonances of the same spin and parity in heavy nuclei verified the surmised Wigner distribution (1956). This was the beginning of Random Matrix Theory (RMT) as it is known today. The impact of RMT in fields as diverse as Number Theory, Condensed Matter Physics and String Theory has far exceeded the expectations of pioneers in the field.

The average (broad) width and (large) spacing of neutron resonances observed in light nuclei can be interpreted in terms of the Nuclear Shell-Model.

For almost four decades, experiments carried out at the Yale Electron Accelerator Laboratory contributed to our knowledge of neutron interactions with nuclei. Before describing some of the key contributions made at the Electron Accelerator Laboratory, a review of the basic theory of neutron resonance reactions is given.

3.2 Neutron Resonance Reactions

The quantities to be discussed in this section may be introduced by considering the problem of s-wave (zero relative orbital angular momentum) scattering of neutrons by a square potential well. The Schrödinger equation for this one-dimensional problem may be written

$$-(\hbar^2/2m)d^2\phi/dr^2 + V\phi = E\phi, \qquad (1)$$

where $V = -V_1$ for $r \leq a$, and $V = 0$ for $r > a$.

The radius of the well is a, and the neutron mass is m. The wave function has to be regular at the origin, so that the wave function inside the well is

$$\phi = A \sin Kr \text{ for } r < a, \qquad (2)$$

where A is an arbitrary coefficient, and K is the wave number inside the well. Outside the well, the wave function is a combination of incoming and outgoing waves:

$\phi = (4\pi v)^{-1/2} \{e^{-ikr} - e^{2i\delta}. e^{ikr}\}$ for r > a, $\qquad$ (3)

where v is the velocity of the neutron, δ is the phase shift, k is the wave number, and $(4\pi v)^{-1/2}$ is a normalizing factor, chosen so that the incoming wave has unit flux. The scattering cross section is related to the phase shift by

$\sigma = (4\pi/k^2) \sin^2\delta.$ $\qquad$ (4)

An alternative way to write the scattering cross section is in terms of a collision function, U, defined as

$U = e^{2i\delta},$ $\qquad$ (5)

in which case, Equn. (4) becomes

$\sigma = (\pi/k^2)[1 - U]^2.$ $\qquad$ (6)

The complicated nature of nuclear forces prevents an exact solution of the collision function; an attempt is therefore made to write U in such a way that it is amenable to simple approximations.

$\qquad$ The scattering cross section given by Equn. (6), has a maximum whenever δ is equal to $(2n + 1) \pi/2$, where n is an integer. For low energies (ka << 1), a maximum will occur for values of a or K such that

$$Ka \approx (2n + 1)\, \pi/2.$$

This condition on Ka for a cross section maximum (a resonance) is a consequence of the fact that $K \gg k$.

The wave function of the scattering cross section resonances at low energies may be interpreted almost as a standing wave. The reflection that occurs, at the edge of the well, keeps the neutron inside for a "long" time. However, the neutron eventually leaves the well so that the wave function is not exactly a standing wave. Nevertheless, a complete set of standing waves X_λ can be constructed of which one, in particular, closely resembles the actual wave function ϕ of the low energy resonance. It can then be shown that, in Fourier expansion of the actual wave function in terms of the standing waves, one term dominates the expansion, and this term corresponds to the one standing wave identified with the low energy resonance. In the nuclear case, the standing waves are the resonance levels of nuclei, and retaining only one term in the expansion leads to the Breit-Wigner formula.

The standing waves, X_λ, may be constructed using both Equn. (1),

$$-(\hbar^2/2m)\, d^2X_\lambda/dr^2 + VX_\lambda = E_\lambda X_\lambda, \qquad (7)$$

and a boundary condition at the square well radius, a,

$$[rdX_\lambda/dr = bX_\lambda]_{r=a} \qquad (8)$$

where b is an arbitrary real number. The boundary condition, given by Equn.(8), coupled with the Schrödinger Equation, Equn. (7), ensures that the X_λ form a complete, orthonormal, set of functions. Therefore, for any value of b, the following Fourier expansion holds

$$\phi = \Sigma_\lambda A_\lambda X_\lambda ; \qquad (9)$$

the coefficients A_λ are given by

$$A_\lambda = \int_{[o,a]} X_\lambda \phi dr. \qquad (10)$$

Using Equns. (7) and (10), it follows that

$$A_\lambda = (1/(E_\lambda - E))(\hbar^2/2ma)X_\lambda(a)\{\phi'(a) - b\phi(a)\}, \qquad (11)$$

where ϕ' means $rd\phi/dr$. Substituting the above value of A_λ in Equn. (8), evaluated at $r = a$, gives

$$\phi(a)=R\{\phi'(a) - b\phi(a)\}, \tag{12}$$

where

$$R = \Sigma_\lambda \, \gamma_\lambda{}^2/(E_\lambda - E), \tag{13}$$

and

$$\gamma_\lambda{}^2 = (\hbar^2/2m)X_\lambda{}^2(a). \tag{14}$$

The R-function, as defined by Equn. (13), is the principal quantity appearing in the formal theory of nuclear reactions.

The collision function, U, can be obtained in terms of R by considering the logarithmic derivative, ϕ'/ϕ at $r = a$, giving

$$\phi'(a)/\phi(a) = (1 + bR)/2. \tag{15}$$

Equn. (3) is then rewritten

$$\phi = I - UO, \tag{16}$$

where $I = (4\pi v)^{-1/2} e^{-ikr}$ and $O = I^*$, are the incoming and outgoing waves, respectively.

By matching the logarithmic derivative of Equn. (15) with Equn. (3) at $r = a$, the collision function becomes

$$U = e^{-2ika}\{(1 + bR + ikaR)/(1 + bR - ikaR) \tag{17}$$

35

To demonstrate that the cross section of Equn. (6) exhibits resonances it is assumed that the energy E is close to a particular E_λ so that the R-function can be represented by

$$R = \gamma_\lambda^2/(E_\lambda - E). \tag{18}$$

Using this value in Equn. (17), and then using the resulting collision function in the cross section expression of Equn. (6), the *Breit-Wigner formula* is obtained

$$\sigma = (\pi/k^2)2\sin ka e^{ika} - [\Gamma_\lambda/\{(E_\lambda - E + \Delta_\lambda) - \Gamma_\lambda/2\}]^2, \tag{19}$$

where $\Gamma_\lambda = 2ka\gamma_\lambda^2$ is the "width" of the level, and $\Delta_\lambda - b\gamma_\lambda^2$ is the "level shift" that moves the maximum of the cross section from E_λ to $E_\lambda + \Delta_\lambda$.

The term $2\sin ka e^{-ika}$ represents potential scattering i.e. the scattering due to a sphere of radius a.

The simplest boundary condition is such that b = 0. The physical resonance states then have zero logarithmic derivative at r = a so that b = 0 makes the standing wave resemble the actual scattering states as closely as possible. This choice of b also makes the level shift zero, so that the characteristic energy of the standing wave then coincides with the energy of the peak cross section. The one-level approximation becomes poor whenever b is so large that the level shift $b\gamma_\lambda^2$ is an appreciable fraction of the spacing between neighboring energies of the standing waves.

Where sharp resonances occur, only a few of the states λ may be involved and contribute to the R-function. The formal theory therefore accounts for the existence of resonances, and allows the cross section to be described in terms of a small set of numbers, namely the reduced widths γ_λ^2 and the resonance energies E_λ. These are the quantities to be measured in neutron interactions.

The extension of the above theory to the general case of nuclear reactions is well understood in principle but is complex in practice. The main points of such a generalization are outlined below:

An excited state of a nucleus can, in general, lose energy (decay), in many different ways, for example, by neutron, proton, a-particle, deuteron emission, γ-radiation etc. These ways are referred to as channels: the simple R-function obtained above for the case of elastic scattering must therefore be replaced by an R-matrix that relates the value of the wave function for a particular channel, at the nuclear surface, to the value and derivative of the actual wave function for all the channels, also at the surface. The rows and columns of the R-matrix then refer to the different channels. The expressions for I, O, U, and b also become channel matrices; Equn. (13) then must be written in matrix notation.

A convenient approximation to the R-function theory is due to Thomas. This approximation is almost always valid in practice, and enables many features of the general theory to be incorporated in the analysis of experimental data without undue numerical complications.

The expression for the total cross section in terms of the collision function U is

$$\sigma_t \quad = (2\pi/k^2)g \, (1 - \mathrm{Re}U),. \tag{20}$$

where g is a statistical weighting factor that depends

on the spins of the incident neutron (1/2) [in units of ħ] and the target (I).

Specifically,

$$g = (2J + 1)/[2(2I + 1)] \qquad (21)$$

in which J is the total angular momentum of the compound state; $[J = I \pm \frac{1}{2}]$ for s-wave neutrons].

At neutron energies below about 1MeV, the most probable processes that occur are the elastic scattering of the neutron and radiative capture, i.e. the absorption of the neutron with subsequent emission of a cascade of gamma rays. (In a number of heavy elements, fission may also occur at these energies).

In the case of s-wave neutron interactions, the collision function is given by:

$$U = \exp(-2ika\}(1 + ikaR)/(1 - ikaR), \qquad (22)$$

(see Equn. (17) with b = 0),

where the R-function may be written

$$R = \Sigma_\lambda \gamma_{\lambda n}^2 /[E_\lambda - E - (1/2)i\Gamma_{\lambda\gamma}], \qquad (23)$$

in which E_λ is the resonance energy of level λ,

$\gamma_{\lambda n}^2$ is the reduced width ($\Gamma_{\gamma n} = 2ka\gamma_{\lambda n}^2$ for s-wave neutrons); $\Gamma_{\lambda n}$ is the neutron width, a is the nuclear radius and k is the neutron wave number ($k = 2\pi/\lambda$). The Thomas R-function, given by Equn. (23), differs from that of Equn. (13) by the term $(\frac{1}{2})i\Gamma_{\lambda\gamma}$ in the denominator. This term is the due to the inclusion of radiative capture in the problem.

Equation (23) is a good approximation provided the partial radiation widths, $\Gamma_{\lambda\gamma i}$, are less than the level spacing, and their amplitudes are random in sign. ($\Gamma_{\lambda\gamma} = \Sigma_i \Gamma_{\lambda\gamma i}$). The collision function given by Equns. (22) and (23) is valid irrespective of whether the total widths Γ_λ ($= \Gamma_{\lambda n} + \Gamma_{\lambda\gamma}$) are greater than or less than the level spacing. This is particularly important for many of the light nuclei in the energy region above several keV.

3.2.1 The Breit - Wigner formula: a useful notation

If there is only one level in the sum over λ in Equn. (23), then the total cross section becomes

$$\sigma_t = \underline{\frac{\sigma_o'}{1 + x^2}} + \tan(2ka)\,\sigma_o'\,\underline{\frac{x}{1 + x^2}} + (4\pi/k^2)\,g\,\sin^2(ka) \tag{24}$$

$$\underset{\text{resonance term}}{|} \qquad \underset{\text{interference term}}{|}$$

The last term represents potential scattering. We have

$$\sigma_o' = (4\pi/k^2) \, g \, (\Gamma_n/\Gamma) \cos(2ka), \qquad (25)$$

$$x = (2/\Gamma) \, (E - E_\lambda), \qquad (26)$$

and

$$\Gamma = \Gamma_n + \Gamma_\gamma, \text{ the total width.} \qquad (27)$$

When $E = E_\lambda$ ($k = k_\lambda$), $\sigma_o' = \sigma_o$, the peak (resonant) total cross section.

For target nuclei with $I \neq 0$, resonances of both spin states are present, in which case the total cross section is obtained by adding the separate contributions. If the contribution from the opposite spin state is added to the last term of Equn. (24), the total potential scattering cross section,

$$\sigma_{pot} = (4\pi/k^2) \sin^2 (ka) \text{ is obtained and Equn. (24) is}$$

then the standard Breit – Wigner formula.

The cumulative effects of distant levels (R^∞, say) on the cross section close to a single level λ may be found as follows

$$R \approx (\Gamma_{\lambda n}/2ka)/(E_\lambda - E - i\Gamma_{\lambda\gamma}/2) + R^\infty \qquad (28)$$

The effect of distant levels is to modify the nuclear radius; using the notation of Feshbach, Porter and Weisskopf, the effective nuclear radius R' is given by:

$$R' = a(1 - R^\infty). \tag{29}$$

3.2.1 A Source of Neutrons

Beginning in the mid-1930's, measurements of the energy-dependence of the total neutron cross section for the interaction of a neutron with a nucleus, have provided essential information in many areas of Pure and Applied Physics.

In 1949, Cockcroft suggested that traveling wave electron linear accelerators could be used to provide intense, pulsed sources of neutrons by making use of (γ, n) and (γ, f, n) photo-reactions. By degrading the primary photo-neutron spectrum with a suitable hydrogenous material, an enhanced flux of low-energy neutrons (< 100keV) could be obtained. The resulting pulsed spectrum of neutrons could then be used for studies of slow neutron interactions with nuclei using time-of-flight methods to determine the neutron energies. Duckworth and Merrison demonstrated the potentialities of this technique by measuring the total yield of neutrons from a Pt-Be target when irradiated with 3.5-MeV electrons from a traveling wave electron accelerator developed by Fry. Although the machine was relatively low power (operating conditions at 3.5 MeV: 1μsec pulse duration, 120 mA peak current, and 200 pulses per sec) a neutron yield of approximately 10^{12} neutrons per sec, during the pulse, was observed. A more powerful accelerator (15 MeV, 1μsec pulse duration, 25 mA peak current at 400 pulses per sec) was installed at Harwell in 1952 for the primary purpose of studying slow neutron interactions with nuclei. At the same time, the electron accelerator at Yale was also being developed as a source of pulsed neutrons for use in the study on low-energy neutron interactions with nuclei.

3.2.2 Measurements of the total neutron cross section

A total neutron cross section is defined as:

$$\frac{\text{The number of events of all types per unit time per nucleus}}{\text{The number of incident neutrons per unit area per unit time}}$$

or, qualitatively, as the effective area that a target nucleus presents to an incoming neutron.

A total neutron cross section $\sigma_T(E)$ is determined by measuring the transmission $T(E)$ of a neutron of definite energy E through an element of uniform thickness. The basic equation is

$$T(E) = \exp\{-n\sigma_T(E)\}$$

where n is the number of nuclei/cm^2 of the element normal to the incident beam.

Although the basic equation is straightforward, there are many technical difficulties to be overcome in making accurate measurements of $T(E)$. They include a) challenges in the production of high fluxes of neutrons in very short intervals of time that are needed to measure the energy E with great precision using the time-of-flight method (the only method available for precise measurements for energies E in the range from a few eV to many MeV); b) corrections for the resolution function of the neutron

44

spectrometer; c) corrections for the Doppler effect due to the relative motions of the incident neutron and target nuclei, and d) determination of the backgrounds that frequently plague neutron experiments. The following example shows resonance structure in the total cross section for the reaction n + U^{238} at energies from 400 – 800eV.

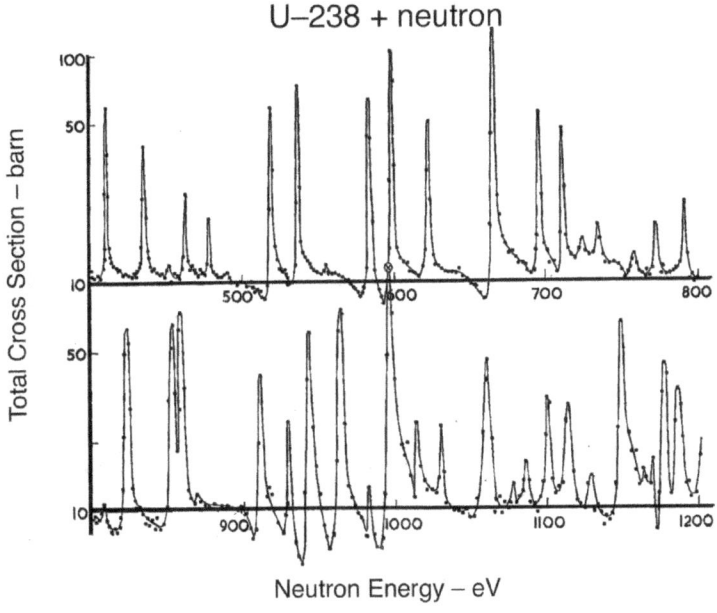

Fig. 9   The observed neutron total cross section for the reaction n + ^{238}U in the energy range 400 to 800 eV.

The asymmetry due to interference between resonant and non-resonant scattering can be clearly seen for the stronger resonances (see Equn. (26)).

A high-resolution study of the total neutron cross section for the reaction ^{23}Na+n is shown in figure 10. An analysis shows that the spin of the resonance is $J = 1/2$, the resonance energy is 2810 $\mp$ 10 eV, and the total width is $\Gamma = 310 \mp 15$ eV.

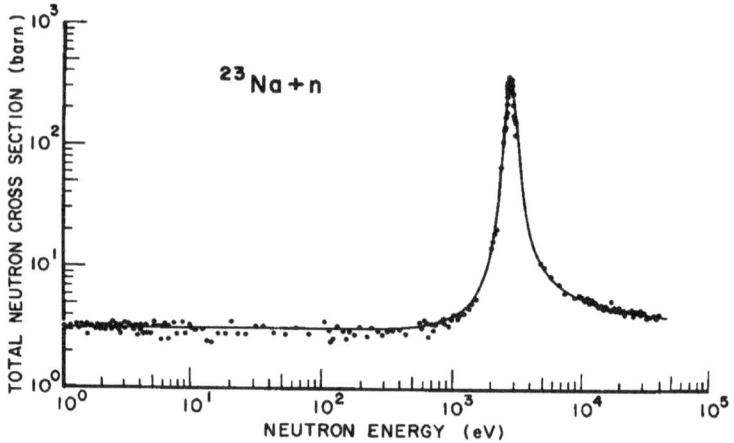

Fig. 10 The measured neutron total cross section for the reaction n + ^{23}Na. The line is a theoretical fit to the data using the above parameters, and a nuclear radius modified by distant levels.

3.3 Neutron Polarization

In the late 1960's, high-resolution neutron polarization studies were carried out at the Laboratory. The program of research, the first of its kind, began with a measurement of the absolute polarization of neutrons scattered from ^{12}C. The experiment involved a neutron double-scattering experiment and nanosecond neutron time-of-flight system. The neutron energies ranged from 2 to 5 MeV, a range in which several resonances of known spin and parity existed. The absolutely calibrated source of polarized neutrons from a ^{12}C scatter was used in a series of experiments to study neutron polarization effects in many light nuclei. The most advanced experiment of this kind involved setting a limit on the neutron polarizability by observing the small angle scattering of polarized neutrons from ^{209}Bi. These experiments are described in later chapters.

3.3.1 Neutron double scattering

The following outline of the theory of neutron double scattering provides a basis for discussing the polarization experiments carried out at Yale over a period of more than 10 years. The spin part of the wave function of a nucleon $|\chi_s>$ may be written in terms of the amplitudes that the spin points "up", a^+, and "down", a^- with respect to a preferred direction, as follows:

$$|\chi_s> = \begin{pmatrix} a^+ \\ a^- \end{pmatrix}$$

If the nucleon spin points in a direction defined by the polar angles (α, β) then:

$$a^+ = \exp\{-i\beta/2\}\cos(\alpha/2)$$

and

$$a^- = \exp\{i\beta/2\}\sin(\alpha/2).$$

The spin part of the wave function after scattering by a spin-0 nucleus, $|\chi_s'>$, is

$$|\chi_s'> = \begin{pmatrix} a'^+ \\ a'^- \end{pmatrix}, \text{ the amplitudes are changed.}$$

The scattering matrix M_1 is defined by the equation

$$M_1 |\chi_s> = |\chi_s'>.$$

If the angular momentum and parity are conserved in the reaction then M_1 must be a scalar. The most general form of M_1 is then

$$M_1 = g_1 \mathbf{1} + h_1 \, \boldsymbol{\sigma} . \mathbf{n}_1$$

where g_1 and h_1 are complex functions of the scattering angle and energy and $\mathbf{1}$, $\boldsymbol{\sigma}$, and $\mathbf{n}_1$ are the unit matrix, Pauli spin matrix and a unit vector normal to the scattering plane, respectively. The direction of $\mathbf{n}_1$ is

$$\mathbf{n}_1 = (\mathbf{k} \times \mathbf{k}')/|\mathbf{k} \times \mathbf{k}'|$$

where $\mathbf{k}$ and $\mathbf{k}'$ are the incident and outgoing momenta of the nucleon.

If the incident beam of nucleons is unpolarized then there are equal numbers of particles with the spin up and spin down, in which case, the differential cross section $(d\sigma_1/d\Omega)_{unpol}$ is found to be

$$(d\sigma_1/d\Omega)_{unpol} = |g_1|^2 + |h_1|^2.$$

It was first shown by Mott in 1929 that the polarization of a beam of electrons could be measured by scattering them a second time and by measuring the left-right asymmetry in their scattering.

In 1946, Schwinger suggested that polarization effects should be observable in neutron-nucleus scattering due to the strong spin-orbit component of the nuclear force. If $\mathbf{n}_2$ is a unit vector normal to the second scattering plane, the angle ϕ between the two planes is given by the scalar product

$$\mathbf{n}_1 \cdot \mathbf{n}_2 = \cos\phi.$$

The polarization produced in the first scattering with respect to the z_2-axis (the z-axis after the second scattering) is

$$P_{1z2} = (|a'^+|^2 - |a'^-|_2)/ (|a'^+|^2 + |a'^-|^2)$$

$$= \{(g_1h_1^* + g_1^*h_1)/(|g_1|^2 + |h_1|^2)\}\cos\phi$$

$$= P_1\cos\phi.$$

We see that the initially unpolarized beam is polarized in the first scattering if $h \neq 0$. This is equivalent to requiring a non-central part in the neutron-nucleus interaction; in the present case, this is provided by the spin-orbit part of the interaction. The polarization vector $\mathbf{P}_1$ is

$$\mathbf{P}_1 = P_1\mathbf{n}_1 ,$$

the polarization is perpendicular to the reaction plane. The spin part of the wave function of the neutron after scattering $|\chi_s'>$, is modified in the scattering from a second spin-0 target nucleus:

$$M_2|\chi_s'> = |\chi_s''>$$

$$= M_2\begin{pmatrix} a'^+ \\ a'^- \end{pmatrix} = \begin{pmatrix} a''^+ \\ a''^- \end{pmatrix}$$

where

$$M_2 = g_2\mathbf{1} + h_2\boldsymbol{\sigma}.\mathbf{n}_2,$$

the matrix associated with the second scattering. The amplitudes after the double scattering are a''^+ and a''^- and the unit vector $\mathbf{n}_2$ is

$$\mathbf{n}_2 = (\mathbf{k}' \times \mathbf{k}'')/|\mathbf{k}' \times \mathbf{k}''|$$

where $\mathbf{k}''$ is the momentum of the neutron after the second scattering.

The differential cross section for the second scattering is

$$d\sigma_2/d\Omega = (|g_2|^2 + |h_2|^2)(1 + \mathbf{P}_1.\mathbf{P}_2)$$

$$= (d\sigma_2/d\Omega)_{unpol} (1 + P_1 P_2 \cos\phi)$$

If the scatterings take place in the same plane then $\phi = 0$ or π, in which case

$$d\sigma_2/d\Omega = (d\sigma_2/d\Omega)_{unpol} (1 \mp P_1 P_2)$$

where the sign depends on whether θ_2 is to the right or left of $\mathbf{k}'$ (having chosen θ_1 on the right of $\mathbf{k}$).

therefore

$$(d\sigma_2{}^R - d\sigma_2{}^L)/(d\sigma_2{}^R + d\sigma_2{}^L) = P_1 P_2$$

or

$$R = (\boldsymbol{R} - 1)/(\boldsymbol{R} + 1) = P_1 P_2$$

where $\boldsymbol{R} = d\sigma_2{}^R/d\sigma_2{}^L$

We see that if $\boldsymbol{R}$ is measured and P_2 is known then P_1 can be determined. Before the experiments at Yale, no absolute values of the polarization were available for P_2 as a function of energy. The Yale studies involved a method in which the first and second scatterers were identical so that $R \simeq P_1^2$. The approximately sign arises because an incident neutron of energy E loses energy on the first scattering (ΔE say) so that the second nucleus is excited at a different energy, $E - \Delta E$. If the polarization P(E) is essentially constant over the range of the energy loss then the true double-scattering relation can be used. Even if the energy loss is appreciable, the following relation is always valid

$$R = P_1(E)P_2(E - \Delta E).$$

3.3.2 Determination of the absolute polarization $P_1(E)$ in the reaction ^{12}C (n, **n**)^{12}C at neutron energies between 2 and 5 MeV.

The experimental arrangement of the true double-scattering experiment is shown in figure 11. Details of the neutron spin-rotation solenoid are described in section 3.3.3.

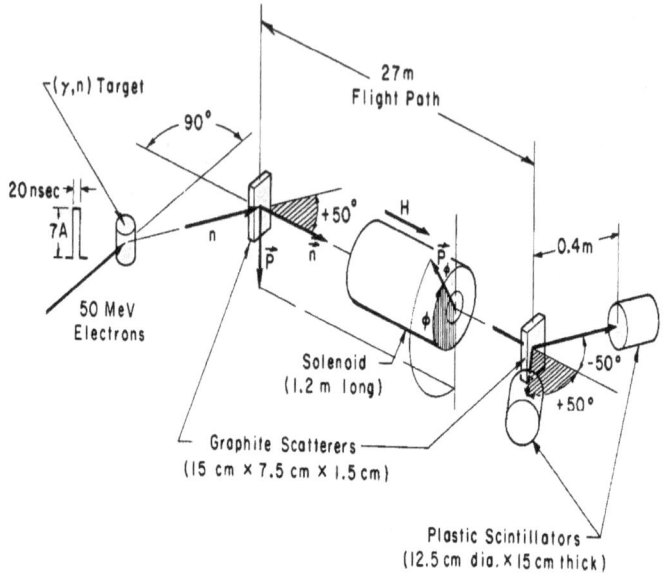

Fig. 11 The arrangement for measuring absolute polarization in n-^{12}C scattering at 50° using a combination of double scattering, a generalized neutron spin-rotation solenoid, and nanosecond time-of-flight spectrometer.

Values of the measured product P(E).P(E − ΔE), and values of the absolute neutron polarization of ^{12}C in the range 2 to 5MeV are shown in figure 12.

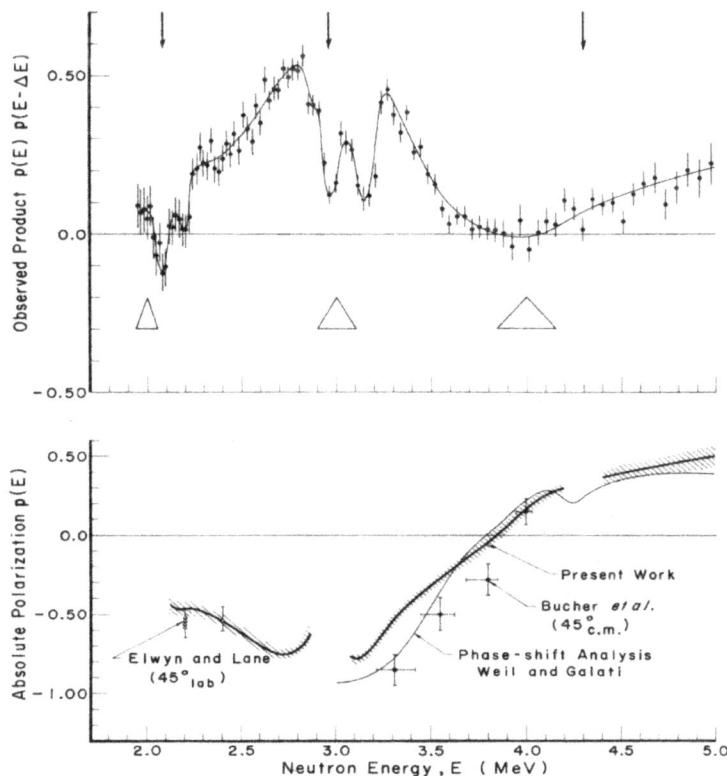

Fig. 12 The measured product P(E).P(E − DE) in the double scattering of neutrons from ^{12}C at 50° and the absolute polarization of the scattered neutrons. The results of earlier measurements are shown for comparison.

The arrows indicate the positions of the 2.08, 2.95 and 3.6MeV resonances in ^{12}C + n. The cross hatching illustrates the small error in the absolute polarization. The energy resolution is indicated by triangles. The neutron energy-loss on scattering through 50° at the 2.08 and 2.95 MeV resonances is clearly seen.

The absolutely calibrated source was used in conjunction with the solenoid and nanosecond time-of-flight spectrometer to carry out a series of polarization studies involving neutron elastic scattering, nuclear photodisintegration and neutron polarizability. Measurements of the ^{4}He(**n**, **n**)^{4}He, ^{6}Li(**n**, **n**)^{6}Li, ^{9}Be(**n**, **n**)^{9}Be, ^{12}C(**n**, **n**)^{12}C and ^{16}O(**n**, **n**)^{16}O scattering reactions, carried out over wide ranges of energy and angle, provided nuclear structure data of very high quality.

3.4 Neutron Spin Rotation

An important part of the polarization studies involved a unique development at Yale of a generalized neutron spin-precession solenoid, shown schematically in the previous figure; the method made it possible to measure, with precision, analyzing powers in the neutron energy range 2 to 10 MeV.

The integrated magnetic field required to precess a neutron of measured energy E_n through an angle of 180° is

$$\int B(z)dz = 2.374 \times 10^5 / E_n \text{ G-cm},$$

and the angle of precession ϕ of a neutron of measured energy $E\phi$ is

$$\phi = \sqrt{(E\pi/E\phi)}.$$

The maximum magnetic field of the solenoid was 5kG and its length was 1.2 meter. The inner diameter was 7.5 cm; the solenoid was an integral part of the neutron collimation system.

The product of the polarization, p, of the source and the analyzing power, A, of the second scatterer is

$$pA = \mp(1 - R_\mp)/(R_\mp - \cos\phi)$$

where $+ \rightarrow$ right detector and

$- \rightarrow$ left detector,

and

$$R_\mp = [N_\mp(H)/N_\mp(0)]\{C(0)/C(H)\}.$$

where $N_\mp(H)$ and $C(H)$ are the corresponding detector count rates and monitor count rates with the field on, and $N_\mp(0)$ and $C(0)$ are the corresponding rates with the field off. The product is thereby obtained independently of the monitor rates.

The effectiveness of the generalized spin-rotation solenoid in covering a wide range of neutron energies with but a single setting of the magnetic field is shown in figure 13. In this example, photoneutrons, scattered from an ^{16}O target at $45°$, passed through the solenoid and then their left/right scattering from a liquid He analyzer was observed at angles of $\mp130°$. Neutron energies ranged from 5 to 10MeV; the angles of precession were from $160°$ at 5MeV to $110°$ at 10MeV. These neutron energies correspond to excitations of the main giant dipole states in ^{16}O.

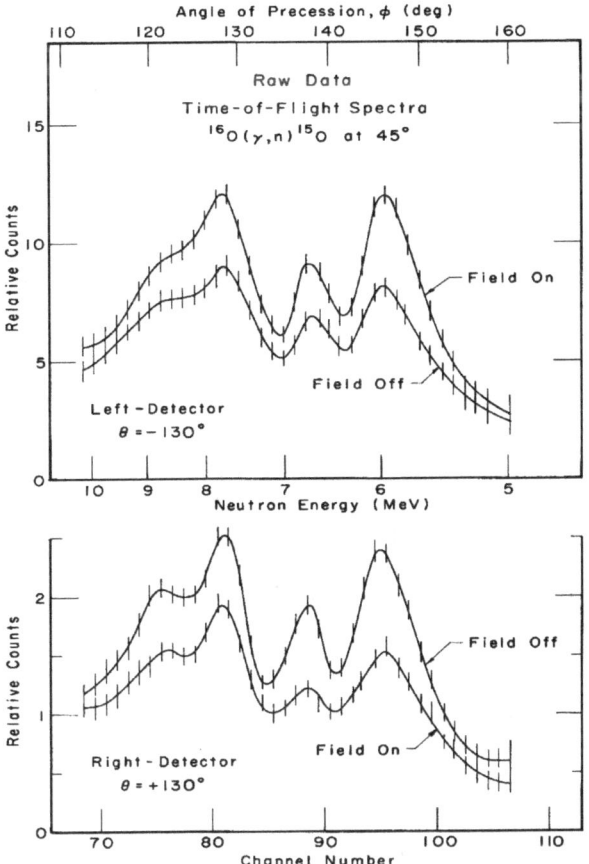

Fig.13 The effectiveness of the generalized neutron spin-rotation solenoid, when set at a definite magnetic field, is seen in the case of scattering of polarized neutrons from the photodisintegration of ^{16}O into detectors at $\mp$ 130.

3.5 Neutron Polarizability

Although the neutron is electrically neutral it has a magnetic moment. The moment implies currents within the neutron, and therefore an internal

59

structure in which the constituents are sources of the currents. As early as 1940, it was found that the magnetic moment of the neutron differs in magnitude and sign from the magnetic moment of the proton. This observation implied a difference in the structure of the neutron and proton – the difference was later explained using the quark model of the nucleons.

The magnetic moment of the neutron can lead to non-nuclear phenomena in the scattering of neutrons in the Coulomb field of the nucleus. In the late 1940's, Schwinger showed that a neutron, moving in the Coulomb field of a nucleus, can undergo a magnetic spin-flip transition and that the amplitude for this interaction, which is electromagnetic in origin, interferes with the amplitude for the purely nuclear interaction between the neutron and the nucleus. He showed that, as a result of the interference, a neutron would undergo a change in its state of polarization, especially at very small, forward angles. The explicit form of the Schwinger polarization is

$$P(\theta) = \{-2\gamma\cot(\theta/2).\mathrm{Im}F(\theta)\}/\{\,|F(\theta)|^2 + \gamma^2\cot^2(\theta/2)\}$$

where $F(\theta)$ is the nuclear scattering amplitude,

$$\gamma = \mu Ze^2/2M_nc^2$$

and $\mu = -1.91$ nuclear magnetons. The other symbols have their usual meanings.

At the small angles of interest $F(\theta) \approx F(0)$ and the optical theorem can be used:

$$\text{Im}F(\theta) = k\sigma_T/4\pi$$

where σ_T is the total nuclear cross section.

In this case, the polarization $P(\theta)$ is a maximum close to $1°$!

With the advent of "meson physics" in the 1950's, a new model postulated that a neutron could be deformed in the nuclear Coulomb field, explicitly:

$$n° \mapsto p^+ + \pi^-$$

in which case, the field interacts with the negatively charged meson or with the proton. The polarizability is intimately associated with the fundamental $p - \pi$ force.

Following the introduction of the quark model of nucleons in the early 1960's, it was clear that a more fundamental model of the neutron's polarizability would involve the complicated interaction between the external electric field and the quark structure. The theoretical problems associated with such a model continue to challenge high-energy theorists to this day. The study of neutron and proton polarizabilities at low energies therefore provides a rare link between Nuclear and Particle physics.

Although the amplitudes associated with the polarizabilty are very small, in favorable cases, it can be determined. The Yale group studied the effect by observing the small angle scattering (3° to 30°) of polarized neutrons from ^{209}Bi in the MeV region. At the time (late 1970's) the results set an improved limit on the value of the electric polarizability coefficient of the neutron. The experiment and the results are discussed in the following pages. The absolutely calibrated source of polarized neutrons was used to interact with a Bi target, 15m from the source. Scattered neutrons were detected in counters placed symmetrically to the left and right of the Bi at angles ranging from 3° to 30°. At angles less than 10° the scattering was dominated by the Schwinger effect. The results were corrected for multiple scattering in the target. The final results are shown in figure 14. The data were compared with theoretical calculations that included the effects of an amplitude associated with the neutron polarizability. Throughout the angular range $< 15°$, evidence showed that the measurements were consistently lower than the values that excluded a polarizability amplitude. A least-squares analysis of the data placed limits on the value of the neutron polarizability coefficient, α_n, namely

$$-1.7 \times 10^{-41} \text{ cm}^3 < \alpha_n < 3.4 \times 10^{-41} \text{ cm}^3.$$

The induced neutron electric dipole moment interacts with an external electric field **E** to give an additional term in the Hamiltonian of the system; this is a second-order effect of the form

$$V_{polarization} = -\alpha_n E^2/2.$$

This was the first limit to be set on the fundamental quantity, α_n using polarization effects in neutron-nucleus scattering.

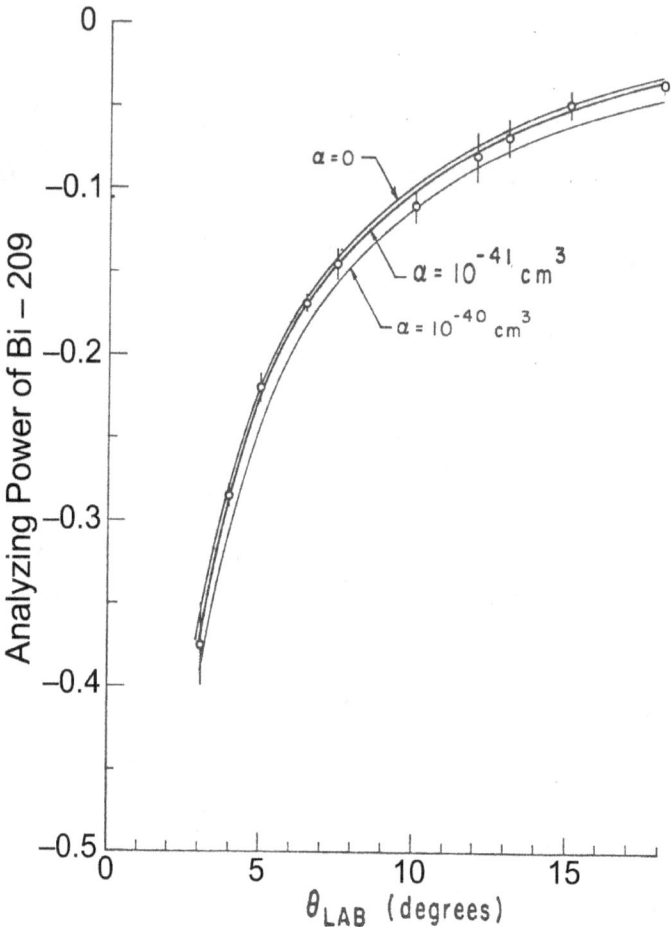

Fig. 14 A comparison between the measured analyzing power of Bi –
209 at a neutron energy of 2.6 MeV, and calculations for different values
of the polarizabilty coefficient, α_n. The angular range is from 3° to 15°.

4. Photonuclear Interactions

4.1 Introduction

In 1948, Baldwin and Klaiber reported the measurement of the cross section for the nuclear absorption of gamma rays with energies in the region of 20 MeV. Goldhaber and Teller interpreted the results in terms of collective oscillations of all protons relative to all neutrons in a given nucleus. The so-called giant electric dipole resonances are a feature of all nuclei. In light nuclei, discrete resonances are observed); the resonances are interpreted in terms of photon absorption into the underlying shell structure of the nuclei.

4.1.1 Preliminaries

The fundamental energy-momentum invariant is:

$$E^2 - (pc)^2 = E^{02}$$

where E is the total energy, p is the relativistic 3-momentum, and E^0 is the rest energy. For a photon, $E^0 = 0$, therefore

$$E_{ph} = P_{ph}c.$$

Using the deBroglie equations

$$E = h\nu \text{ and } p = h/\lambda,$$

we obtain for a photon:

$$E_{ph} = hc/\lambda_{ph}, \text{ which leads to}$$

$$\lambda_{ph}/2\pi \approx \{200/E_{ph}(Mev)\} \text{ fm } (10^{-13}cm).$$

The area of a nucleus is given in terms of the transmission of a well-collimated beam of low-energy neutrons through a sample containing the nuclei:

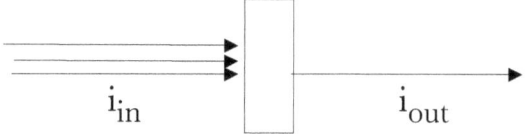

$$\text{transmission, } T = i_{out}/i_{in} = \exp\{-n\sigma_t\}$$

where n is the number of nuclei /cm² of the sample, and σ_t is the cross section for the interaction.

Measurements give $\sigma_t \approx 10 \cdot 10^{-24}$ cm² (10 barns).

For heavy nuclei (A > 200), σ_t is related to the nuclear radius: $\sigma_t = 4\pi r_{nuc}^2$; for heavy nuclei, $r_{nuc} \approx 10$ fm.

We see, therefore, that to study an object of nuclear size, we need a normalized photon wavelength $\lambda/2\pi = \lambdabar \approx 10$ fm and therefore a photon energy of about 20MeV.

The field of a photon can be written in terms of a series of multipoles, each one carrying a definite angular momentum and parity. The most important multipole for photonuclear interactions in the MeV-region involves the electric dipole (E1) component of the field ($\lambda_{ph} > r_{nuc}$). It transfers one unit of angular momentum and odd parity to the absorbing nucleus, so that a nucleus with a spin and parity 0^+ is excited into a state with spin and parity 1.

In the E1-approximation, the electric field is essentially constant over the nucleus. Migdal(1944), and Goldhaber and Teller(1948), first discussed the theoretical aspects of E1 photonuclear interactions.

Their model suggested that the photon electric field displaces all the protons (+ charges) relative to the neutrons (neutral), thereby setting up a nuclear dipole moment. The very strong inter-nucleon forces (acting between pairs) produce a restoring force. Goldhaber and Teller calculated that the protons and neutrons would oscillate, as two groups, at a resonant frequency corresponding to photon energies between 10 and 25 MeV (the exact frequency depends on the size (and therefore the mass number) of the nucleus).

In 1962, Professor Firk, working in England, carried out the first experiments on the (γ, n) reaction in light nuclei with an energy resolution of 10 keV. With such high resolution, he was able to show that the single giant resonance of the Goldhaber-Teller model is often split into hundreds (sometimes thousands) of closely spaced resonances. These studies were continued when he joined the Yale Electron Accelerator group in 1965. The energies of the absorbed photons were deduced from the relativistic kinematics of the reaction. In the late

1960's, comparisons between high-resolution photoneutron experiments, carried out at Yale, and the results of high-resolution studies of γ, p reactions gave what is, to this day, some of the best evidence for the charge-independence of the nuclear force.

4.2. Differential Polarization of Photoneutrons and Differential Cross Section of Photoprotons from ^{16}O

The work described here was the culmination of a program of developments by members of the Laboratory that lasted 15 years; the program dealt with the challenging task of measuring the polarization of neutrons (including photoneutrons) with high resolution at energies ranging from 1 to 50 MeV. The key technical developments made during that period of time have been discussed in previous chapters. Here, results of a detailed study of the differential polarization of photoneutrons, emitted from ^{16}O throughout the region of the giant dipole states, are given.

The generalized neutron spin precession solenoid, first developed in the Laboratory, was combined with a nanosecond time-of-flight spectrometer and left-right scattering from a liquid helium (or graphite) polarization analyzer to determine the polarization of photoneutrons, $p_y(\theta)$, scattered through an angle θ, from ^{16}O. The neutron energies ranged from 1 to 15 MeV. A view of the liquid helium neutron polarimeter is shown

Fig 15. The liquid helium neutron polarimeter, hanging from its ceiling mount. The two neutron detectors (in cylindrical cases) detected neutrons scattered to the right and left of the liquid helium at angles of 135 ^{O}. In this figure, the polarimeter is at the end of a 30-meter flight tube; a resolution of 10 keV was achieved when studying 1 MeV neutrons, a factor of about 100 times better than all previous studies of neutron polarization in nuclear photodisintegration.

4.2.1 Determination of the differential polarization coefficients

The magnitude of the differential polarization $|dP/d\Omega|$ is related to the measured product $p_y(\theta)\sigma(\theta)$ where $\sigma(\theta)$ is the differential cross section. In the case of the reaction $^{16}O(\gamma, n_0)^{15}O$, in the E1–E2 approximation, we have

$$|dP/d\Omega| = (1/2)p_y(\theta)\sigma(\theta) = \sum_{k=1}^{4} B_k P_k^1(\cos\theta) .$$

where the coefficients B_1–B_4 that multiply the associated Legendre polynomials $P_k^1(\cos\theta)$ can be expanded in terms of sine-functions as follows

$$(1/2)p_y(\theta)\sigma(\theta) = -B_1[0.86603(\sin\theta)] - B_2[0.96825(\sin2\theta)]$$
$$-B_3[0.20253(5\sin3\theta+\sin\theta)]$$
$$- B_4[0.14823(7\sin4\theta + 2\sin2\theta)].$$

A least-squares analysis of the products $(1/2)p_y(\theta)\sigma(\theta)$ gives the optimum values of B_1, B_2, B_3, and B_4. The following equation relates the B_k-coefficients to the amplitudes b_k:

$$dP/d\Omega = \lambda_\gamma^2(\sqrt{3}/8)\{\sum_{k=1}^{4} b_k P_k^1(\cos\theta)\} = \sum_{k=1}^{4} B_k P_k^1(\cos\theta)$$

where

$$b_1 = 1.581a_s a_p \sin\delta_{sp} - 0.894a_d a_p \sin\delta_{dp} - 1.643a_d a_f \sin\delta_{df}$$

$b_2 = -0.949a_s a_d \sin\delta_{sd} + 0.652a_p a_f \sin\delta_{pf}$

$\dots b_3 = 1.380a_s a_f \sin\delta_{sf} + 1.434a_d a_p \sin\delta_{dp} + 0.195a_d a_f \sin\delta_{df}$

$\dots b_4 = -2.130a_p a_f \sin\delta_{pf}$.

The quantities $\delta_{sp} = \phi_s - \phi_p$, $\delta_{dp} = \phi_d - \phi_p$, $\dots$ are the phase differences between the real phase shifts ϕ_s, ϕ_p, ϕ_d, ... The coefficient b_4 contains amplitudes associated only with E2-absorption.

The phase shifts occur as phase differences and therefore one of the phase shifts can be chosen arbitrarily. Seven variables then remain to be determined. A combination of the results from the measurements of the angular distribution coefficients (see next paragraph) and the Yale measurement of the differential polarization coefficients were used to give a complete solution to the problem.

The differential cross section $\sigma(\theta)$ in the E1–E2 approximation is

$$\sigma(\theta) = (\lambdabar_\gamma^2/8)\sum_{k=0}^{4} a_k P_k(\cos\theta) = \sum_{k=0}^{4} A_k P_k(\cos\theta),$$

where the coefficients are

$$a_0 = 3(a_s^2 + a_d^2) + 5(a_p^2 + a_f^2),$$

$$a_1 = 9.487 a_s a_p \cos\delta_{sp} + 9.859 a_d a_f \cos\delta_{df} - 1.342 a_d a_p \cos\delta_{dp},$$

$$a_2 = 4.243 a_s a_d \cos\delta_{sd} - 1.5 a_d^2 + 2.5 a_p^2 + 2.857 a_f^2$$
$$- 1.75 a_p a_f \cos\delta_{pf},$$

$$a_3 = 7.746 a_s a_f \cos\delta_{sf} - 4.382 a_d a_f \cos\delta_{df}$$
$$+ 8.05 a_d a_p \cos\delta_{dp},$$

and

$$a_4 = 13.997 a_p a_f \cos\delta_{pf} - 2.857 a_f^2.$$

(Note that $A_0 = \sigma_{total}/4\pi$ (mb.sr^{-1}) where σ_{total} is the total (integrated-over-angles) cross section).

4.2.2. The E1 reaction amplitudes and phases.

The complexity of the equations for the differential polarization and the differential cross section, and errors in the coefficients a_i, b_i, make it necessary to adopt an iterative method of solution that relies upon the known dominance of the E1-amplitudes, the relative smallness of the E2- and the negligibly small M1-amplitudes above 20 MeV. Furthermore, earlier polarization measurements carried out at the Laboratory indicated that the ratio of the s-wave to d-wave amplitudes, a_s/a_d, is typically less than 1/4. The analysis therefore begins by solving the following three equations that involve terms of the forms a_s^2, a_d^2, and $a_s a_d$, and the phase difference δ_{sd}:

$$a_0 = 3a_s^2 + 3a_d^2$$

$$a_2 = 4.243 a_s a_d \cos\delta_{sd} - 1.5 a_d^2$$

$$b_2 = -0.949 a_s a_d \sin d_{sd}.$$

Values for the three unknowns a_s, a_d, and ϕ_d ($\phi_s = 0$), are thereby obtained. Results of the final analyses of the high-resolution polarized photoneutron experiments for the reaction $^{16}O(\gamma, n_0)^{15}O$ are shown in the following figures:

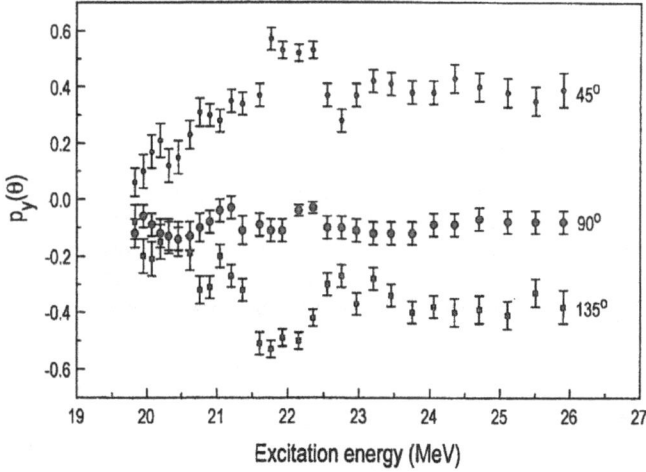

Fig.16 Measured polarizations for the reaction $^{16}O(\gamma, n_0)$ ^{15}O at 45°, 90° and135°.

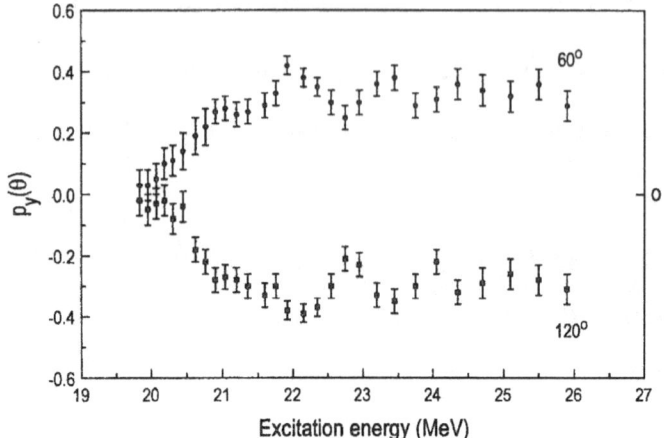

Fig. 17 Measured polarizations for the reaction $^{16}O(\gamma, n_0)$ ^{15}O at 60° and 120°

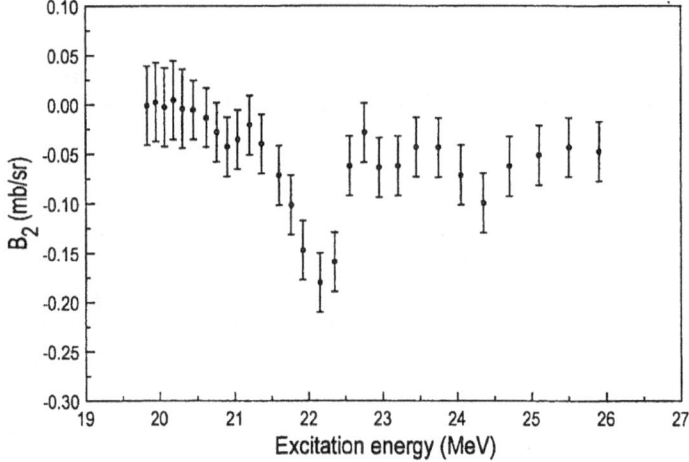

Fig. 18 The B_2 coefficient obtained from an analysis of the ^{16}O (γ, n_0) ^{15}O data.

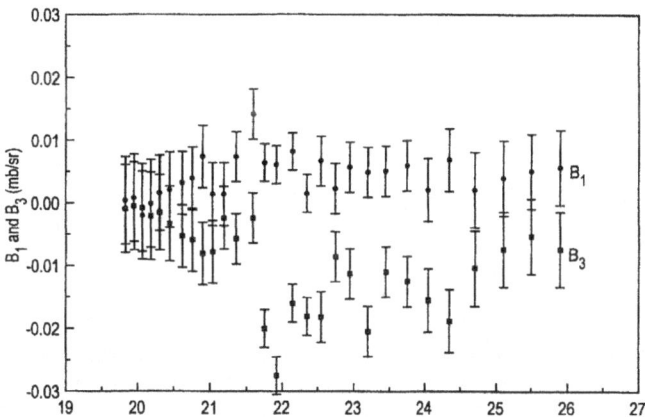

Fig. 19 B_1 and B_3 coefficients obtained from an analysis of the $^{16}O(\gamma, n_0)$ ^{15}O data.

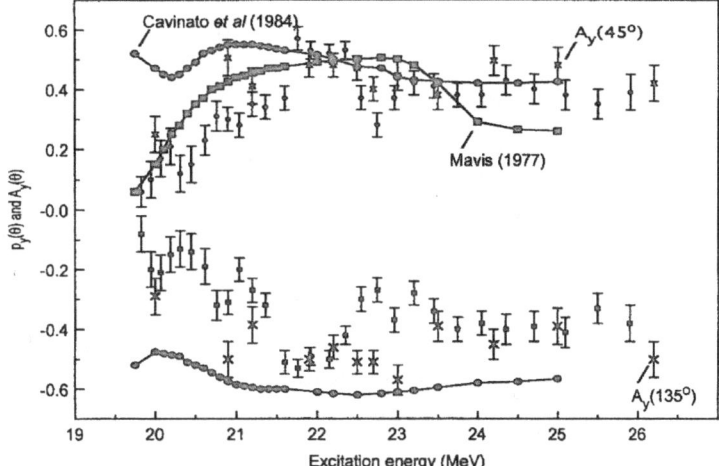

Fig. 20 Comparisons between the measured values of $p_y(\theta)$ at 45° and 135°, the measured values of $A_y(\theta)$ from photoproton data, and theoretical calculations.

The theories do not account for the observed structure throughout the energy region and, in the case of the 135° measurements they do not reproduce the observed magnitudes.

4.2.3 Measurement of the differential cross section
of the reaction $^{16}O(\gamma, p_0)^{15}N$

Large volume Li-drifted silicon counters were constructed and placed at select angles in a scattering chamber. A target, ^{16}O, was irradiated with bremsstrahlung beams of differing end points, with maximum energies chosen to provide a measurement of the ground state cross section as a function of excitation energy and angle. The value of the ^{16}O cross section was normalized to the known deuterium cross section. The data were analyzed to give the Legendre coefficients A_0 through A_3; the results are shown in figure 21.

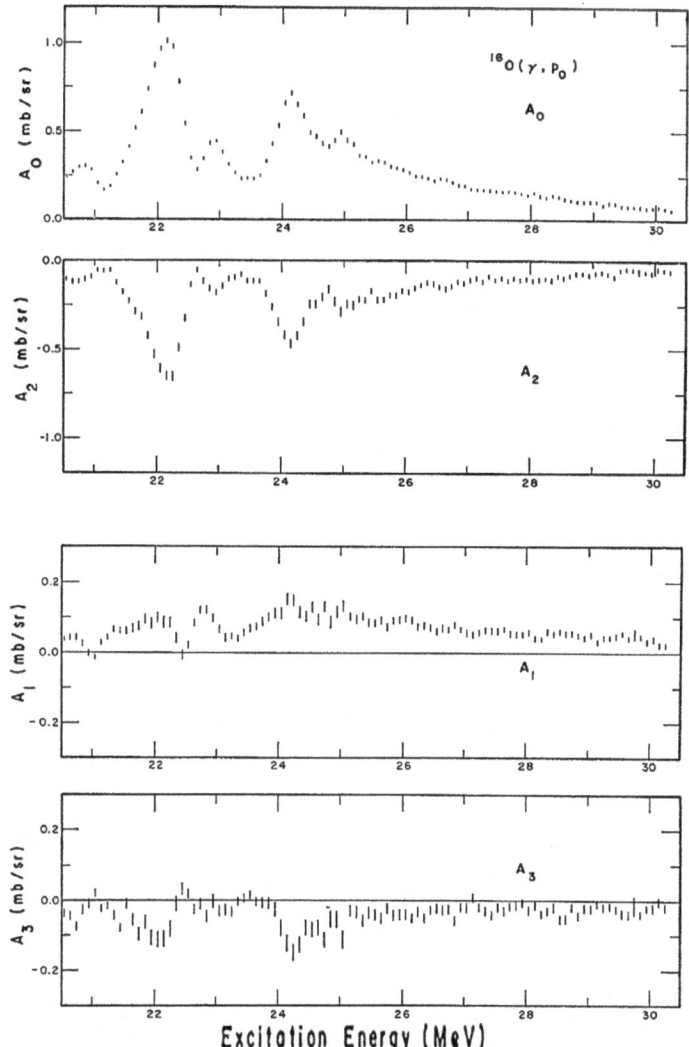

Fig. 21 The angular distribution coefficients for $^{16}O(\gamma, p_0)^{15}N$ in the giant resonance region.

4.3 Isospin Mixing in Photoreactions

In a self-conjugate nucleus, the ratio of photoproton to photoneutron cross sections, leading to mirror levels in the residual nuclei, is related to the amplitudes of isospin componenets. For isolated states with single exit channels, Barker and Mann deduced the expression

$$\sigma(\gamma, p)/\sigma(\gamma, n) \approx (P_p/P_n)\,|(\alpha_1 + \alpha_0)/(\alpha_1 - \alpha_0)|^2$$

where P_p and P_n are the proton and neutron penetrabilities respectively, and α_0 and α_1 are the amplitudes of the isospin $T = 0$ and $T = 1$ components of the excited state wave function. Barker and Mann's expression assumes that the transition probabilities may be expressed in terms of reduced widths and that the angular distributions of the proton and neutron channels are identical. The high resolution measurement of the $^{16}O(\gamma, n_o)^{15}O$ reaction, made in the Laboratory, was compared with the $^{16}O(\gamma, p_o)^{15}N$ measurement to give values of the α_o

and α_1 as a function of excitation energy from 17 to 40 MeV. The results are shown in figure 22.

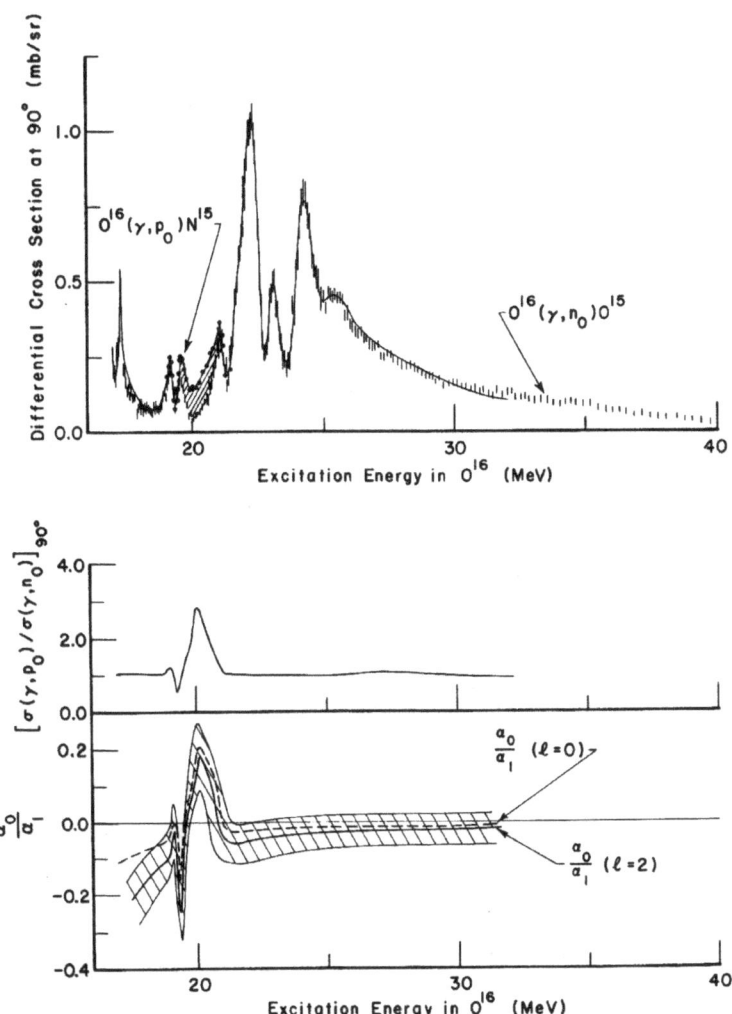

Fig. 22 Isospin mixing in the ground state photoneutron and photoproton 90° differential cross sections of ^{16}O in the giant dipole resonance region.

4.4 Isospin Splitting of Giant Dipole Resonances.

As early as 1955, Morinaga had suggested that, in certain nuclei, the giant dipole state could be split into two components, characterized by the isospin quantum numbers $T = 1$ and $T = 2$. Speculative arguments were made over the years that such splitting was present in observed measurements, particularly in inelastic electron scattering experiments. However, definitive results were needed to establish such a splitting. *In 1970, clear evidence was obtained at the Yale Laboratory for the isospin splitting of the giant resonance in ^{26}Mg.* The experiment involved studying the energy spectra of photoneutrons emitted from ^{26}Mg with carefully chosen end-points for the bremsstrahlung spectra used to excite the nucleus. It was found that states centered around an excitation energy of 17 MeV decayed strongly to the ground or low-lying states of ^{25}Mg ($T = \frac{1}{2}$) whereas those around 22MeV decayed to the $T = 3/2$ states in ^{25}Mg:

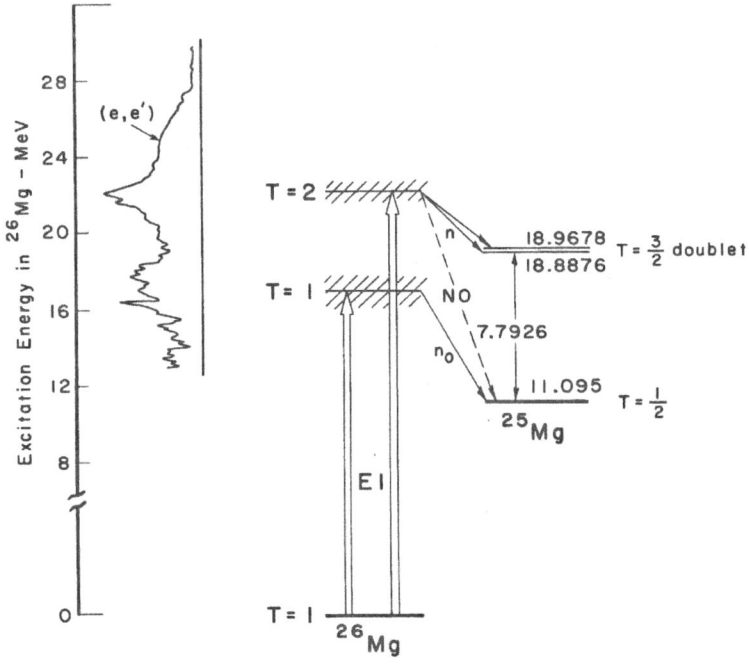

Fig. 23   Energy level diagram showing the locations of the dipole states in ^{26}Mg. The transitions from the expected T=2 states to the T=3/2 doublet in ^{25}Mg around 7.8 MeV are indicated. The results of an e, e' experiment are also shown.

Energy spectra were measured with bremsstrahlung end-points of 18.9, 21.4, 23.1, and 27.5 MeV, and difference spectra were obtained; here are two spectra and their difference:

83

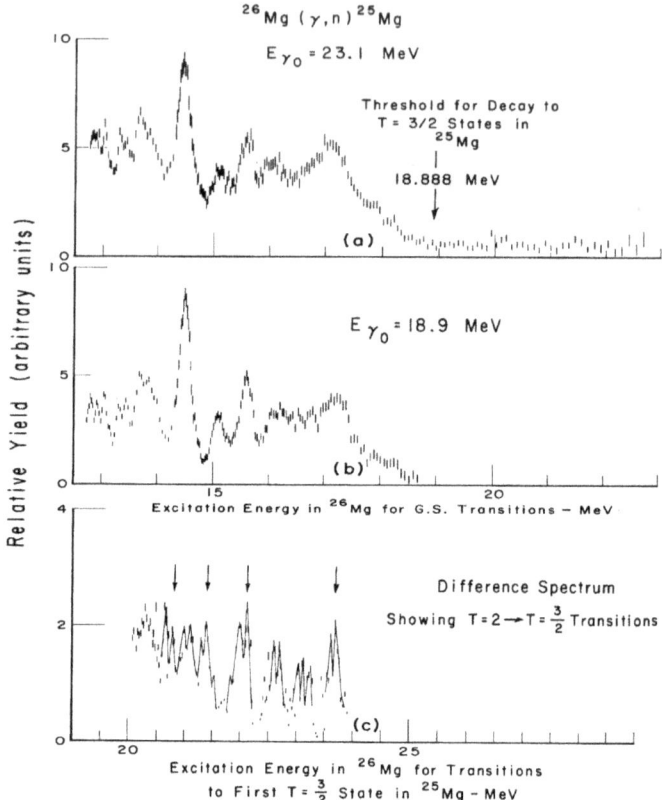

Fig. 24   The energy spectra of photoneutrons from ^{26}Mg(γ, n) ^{25}Mg for bremssrahlung end-points of 23.1 and 18.9 MeV [(a) and (b)}. The difference spectrum is shown in (c); the arrows show doublets associated with transitions to the (T = 3/2) 7.7926 and 7.7829 MeV states in ^{25}Mg. They correspond to T=2 states in ^{26}Mg at excitations of 22.70, 22.13, 21.37, 21.16, 20.77 and 20.37 MeV. The T = 1 states center at an excitation energy $\approx$ 17 MeV.

4.5 Photodisintegration of the Deuteron

From the beginning of Nuclear Physics in the early 1930's studies of the interaction of real photons with nucleons and nuclei at photon energies below about 50 MeV have been an important part of the field. In the standard textbook *Subtomic Physics* by Frauenfelder and Henley, the authors remark "Another celebrated case (of study) is the photodisintegration of the deuteron:

$$\gamma d \rightarrow np$$

which was discovered in 1934 by Chadwick and Goldhaber and used by them for a measurement of the neutron mass". In Chapter 4, the results of the Yale studies of the interaction of photons with nuclei in the energy range from $10 - 30 MeV$ were discussed. Here, samples of the many experiments on the $\gamma d \rightarrow np$ reaction, carried out in the Laboratory over the decades, are given.

4.5.1 Angular distribution of photoprotons from the deuteron.

Professor Schultz and his graduate student, Barry Weissman, measured the differential cross section of photoprotons over a wide range of angles and at many excitation energies (27 – 55 MeV). Their experiment involved the local manufacture of high-quality lithium-drifted silicon proton detectors of large volume in order to cover a wide range of photoproton energies. Their results are shown in figure 25 where they are compared with other measurements, and with the theoretical predictions by Partovi. It is interesting to note that the theoretical calculations represented the most fundamental of any carried out in Nuclear Physics up to that time. They involved applications of the best known models of the nucleon-nucleon interaction.

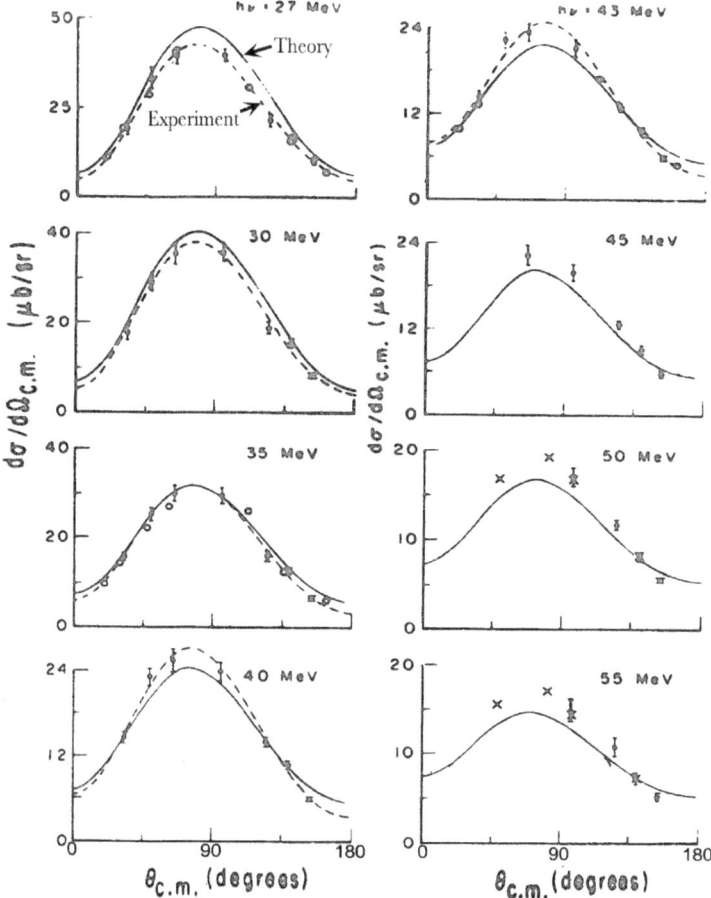

Fig. 25 The measured differential cross sections for the d(γ, p)n reaction (closed circles) and open circles (another laboratory) . The dashed lines are fits to the data using a standard Legendre expansion. The solid line represents the best available theoretical predictions.

The fitted curves are of the standard form

$$\sigma(\theta) = \Sigma_{l=0,n} A_l P_l(\cos\theta)$$

The angular distribution coefficients A_l obtained from the fits ranged from A_0 to A_4. This experiment took 3 years to set up and to complete; the accelerator beamtime used was 1500 hours.

4.5.2 The differential polarization of photoneutrons
        from the reaction d(γ, n)p.

Professor Firk and his graduate student Larry Drooks searched for evidence of meson effects in the d(γ, n)p reaction at photon energies between 6 and 15MeV and at angles of 60°, 90°, and 120°. The polarization of the photoneutrons was measured using the absolutely calibrated analyzer, ^{12}C, placed 25m from the liquid deuterium target. The neutron energies were measured using a nanosecond time-of-flight spectrometer. This was a challenging experiment carried out over a period of 18 months. The final results, obtained by taking the mean values of numerous independent experiments were:

Mean energy range (MeV)	Mean polarization	Lab °
9.85	−0.111 + 0.024	60
9.32	−0.112 + 0.013	90
9.83	−0.129 + 0.013	121

The value of −0.112 + 0.013 at 90° is 2 standard deviations less than the classical (non-meson) calculation whereas it is within ½ standard deviation of a calculation that includes the effect of meson currents. This tantalizing result requires further study, 40 years on.

4.6 A New Computer System

In 1966, a grant of $250,000 was received from the Government to install a coupled two-computer system at the Yale Electron Accelerator laboratory. This ushered in a new era in doing experimental physics and on-line analysis of data. Also, the system permitted control of the accelerator itself. Accelerator control was new not only at Yale but also throughout the world of Accelerator Physics.

The system was designed by Professor Firk, in conjunction with experts from the Digital Equipment Corporation, then a leading manufacturer of systems for serious computing, and of unique systems of the kind required at Yale.

The main features of the system were the use of two coupled computers, one was a small PDP-8 for control of electron scattering experiments and the control of the energy and other parameters of the operating accelerator The other, a PDP-7, was for use as a 4096-channel nanosecond neutron time-of-flight spectrometer, capable of use as eight independent 512-channel spectrometers, and for on-line data analysis. The coupled system presented a challenge for Graduate student programmers because the basic word-length of the PDP-8 was only 12 bits, whereas for the PDP-7 it was 18 bits.

Initially, the programs were developed using paper tape, and data output was recorded on large magnetic tapes – how things have changed in the last 50 years!

5. Academic Staff

5.1 Professors

Howard L Schultz, Director 1946-1976

Charles K Bockelman, Associate Director 1960-68

Frank W K Firk, Assoc. Director 1968-76, Director 1976-84.

Assistant Prof. Gerald A Peterson/ Prof. Univ. of
Massachusetts at Amherst

5.2 Visiting Faculty

Frank D Brooks/Prof. University of Cape Town

Barry L Berman/Prof. The George Washington Univ.

5.3 Research Staff with Later Appointments

James E Draper – Professor, Univ. California, Irvine

Thomas H Schucan – Professor, Univ. Basle, Switzerland

John E E Baglin – Director, Research Div. IBM San Jose

Maxwell N Thomson – Professor, Univ. of Melbourne AU

Robert O Owens – Professor, Univ. of Glasgow, Scotland

Horst Theissen – Senior Scientist, Darmstadt, Germany

James O'Connell – National Bureau of Standards

5.4 Visiting Research Staff

Jonas Alster – Northeastern Univ.

Jean Bellicard – Senior Scientist, Saclay, Paris, France

Benson Chertok – American Univ.

William Dodge – National Bureau of Standards

James Stewart – Univ. of New Haven

5.5 Post-doctoral Fellows

Michel A Duguay –Professor, Laval Univ., Canada

Chung-Pao Wu – Senior Scientist, RCA, Princeton NJ

Ravinder Nath –Professor, School of Medicine, Yale Univ.

Roy J Holt – Prof., Univ. of Illinois, Director Intermediate
Energy Physics Div. Argonne National Lab.

Lawrence Cardman – Prof. Univ. of Illinois, Director of
Research, The Thomas Jefferson Laboratory VA.

5.6 Technical and Support Staff

A list of technical staff appointments in the Laboratory is
given in Chapter 2.

The list of visiting scientists, often from far
and wide, and the successes of Graduates of the
Laboratory, attest to the vitality of the laboratory in
the 1960'and 70's.

6. Theses

The following list shows the wide range of research carried out by Ph D students in the Laboratory over the decades. Included in the list are 8 theses in the field of electron scattering, a topic that is not dealt with in the present book. Their work was, however, often closely connected with the research in neutron and photoneutron physics reported here.

Year	Name	Thesis Title	Advisor
1948	C L Clarke	Electron Ballistics in a Cavity Linear Accelerator	H L Schultz
1948	B G Farley	Measurement of Nuclear Particle Velocities by a Time Coincidence Method	H L Schultz
1948	G A Kolstad	A Linear Accelerator for the Production of High Intensity Gamma Rays and Neutrons	H L Schultz
1848	J A Lockwood	Development of a Linear Electron Accelerator III: Design and Development of Accelerator Column	H L Schultz
1948	R L McCarthy	Development of a Linear Electron Accelerator II: Radio frequency Power Generation	H L Schultz

Year	Name	Thesis Title	Advisor
1948	P J Rice Jr	Development of a Linear Electron Accelerator I: Measurement Techniques and Applications	H L Schultz
1948	F W van Name Jr	Measurement of the Lifetime of a Radioactive Element of very short period	H L Schultz
1950	D Binder	Delayed Coincidence Counting Measurements	H L Schultz
1950	W J MacIntyre	Measurements of Short-lived Isomers	H L Schultz
1952	M S Malkin	Electron Mobilities in Liquid and Solid Argon and Liquid Helium	H L Schultz
1953	T C Engelder	A Search for Shot-lived Nuclear Isomers	H L Schultz
1953	T D Strickler	Large-Angle Scattering of Co^{60} Gamma Rays	H L Schultz
1954	C W Hoover Jr	Secondary Electron Resonance Discharge Mechanism and Transient Response of Cavity Resonator Accelerating Sections	H L Schultz
1955	J G Carver	Slow Neutron Resonances in Indium and Tungsten	H L Schultz
1955	T F Godlove	Slow neutron Resonances in Tantalum and Tellurium	H L Schultz
1956	R G Bennett	Gamma Rays Following Single-Level Neutron Capture	H L Schultz

Year	Name	Thesis Title	Advisor
1957	C A Fenstermacher	Gamma Ray Transitions in Heavy Nuclei Excited by Resonant Neutron Capture	H L Schultz
1957	A E Walters	Resonance Neutron Capture Gamma-Rays from Heavy Nuclei	H L Schultz
1958	L Rosler	Energy Level and Decay Scheme of Sm ^{150}	H L Schultz
1959	T E Springer	Multiplicity of Gamma-Rays from Resonance Neutron Capture in the range A = 110 to 198	H L Schultz
1960	A A Fleischer	Two-Step Gamma Ray Cascades Following Slow Neutron Capture in Cl, V, Ti and Mn	J E Draper
1962	C O Bostrom	Thermal Neutron Capture Gamma Rays	J E Draper
1964	O A Wasson	Direct Neutron Capture in Gold	J E Draper
1965	W J Alston III	The Photoproton Polarization from C^{12}	H L Schultz
1965	J R Stewart	Photodisintegration of Helium-3	J O'Connell
1965	R H Hilberg	The Polarization of Photoprotons from Beryllium	H L Schultz
1965	K J Wetzel	Investigation of the Neutron Capture Mechanism in Gold	C K Bockelman
1966	M. Duguay	Inelastic Scattering of Electrons from Even Nickel Isotopes	C K Bockelman
1966	B J Liles	The Angular Distribution of Photoptotons from Deuterium	H L Schultz

Year	Name	Thesis Title	Advisor
1966	R C Morrison	The Photoproton Cross Sections of Oxygen-16 and Carbon-12	H L Schultz
1967	J F Ziegler	Collective Excitation in Pb^{206}, Pb^{207}, Pb^{208} and Bi^{209} by Inelastic Electron Scattering	G A Peterson
1968	T H Curtis	Quadrupole and Octupole Excitations of the Even Tin Isotopes	C K Bockelman
1969	R A Eisenstein	Studies of Some Even Calcium Isotopes by Inelastic Electron Scattering	C K Bockelman
1969	C-P Wu	A Study of the Giant Dipole States in Some Light Nuclei	F W K Firk
1970	D W Madsen	A Study of Some Even Isotopes of Neodymium by Electron Scattering	C K Bockelman
1970	B Weissman	Photodisintegration of the Deuteron: Absolute Cross Section and Angular Distribution	H L Schultz
1971	G W Cole Jr	The Polarization of Photoneutrons from Oxygen-16	F W K Firk
1971	M G Mustafa	Photonuclear Reactions in Some Light Nuclei	H L Schultz
1972	E J Bentz Jr	Mechanisms of the Giant Resonances	J E E Baglin
1972	L S Cardman	Electron Scattering Studies of Some Even Isotopes of Samarium	C K Bockelman
1972	D Kalinsky	Electron Scattering Studies of W^{184} and W^{186} Nuclei	C K Bockelman
1972	R Nath	The Polarization of Photoneutrons from Deuterium	F W K Firk

Year	Name	Thesis Title	Advisor
1973	R J Holt	The Absolute Polarization of Neutrons from n-C^{12} Scattering	F W K Firk
1973	J Legg	A Study of the Low-Lying States of $Ba^{148,150,152}$ and Ba^{138} by Electron Scattering	C K Bockelman
1973	R Y Yen	Measurement of Reduced Radiative Transition Probabilities by Electron Scattering	C K Bockelman
1974	G T Hickey	The Polarization of neutrons in n-O^{16} Scattering	F W K Firk
1975	J E Bond	The Neutron-Alpha Particle Interaction: A Study of Polarization Effects up to 6MeV	F W K Firk
1976	L J Drooks	The Polarization of the Neutrons Produced by the Photo-Disintegration of Deuterium	F W K Firk
1979	Y H Chiu	The Polarization and Angular Distribution of Fast Neutrons Scattered Elastically from Lithium-6	F W K Firk
1981	M Ahmed	Nuclear and Non-Nuclear Polarization Effects in the Elastic Scattering of Fast Neutrons from Bismuth-209	F W K Firk
1983	J W Kruk	A Study of the Electric Polarizability of the Neutron by Low Energy Resonant Scattering	F W K Firk
1984	A Vazquez	Experimental and Theoretical Studies of the Giant Resonances in Nuclei	F W K Firk

7. Some Publications

The following selection from the several hundred publications that were produced by members of the Laboratory over a period of four decades provides details of notable publications during those decades.

1. *Cavity Accelerator for Electrons*,

   H L Schultz, R Beringer, C L Clarke,

   J A Lockwood, R L McCarthy, C A Montgomery,

   P J Rice and W W Watson.

   Phys. Rev. **72**, 346 (1947).

2. *The Yale Linear Accelerator*

   H L Schultz and W G Wadey,

   Review of Scientific Instruments, **21**, 383 (1951).

3. *Electron Mobilities in Liquid Argon*

   M S Malkin and H L Schultz.

   Phys. Rev. **83**, 1051 (1951).

4. *Variations in Spectra of Neutron Capture*
   Gamma Rays in Indium

   J E Draper, C A Fenstermacher and H L Schultz

   Phys. Rev. **111**, 906 (1958)

5. *Thermal and Resonance Neutron Capture in Copper, Nickel and Manganese*
   O A Wasson and J E Draper,
   Phys. Rev., **137**, B1175 (1965)

6. *Fast Neutron Spectroscopy Using the Yale Electron Linac and a New Nanosecond Time-of Flight System*
   F W K Firk, Nucl. Instr. and Meths. **43**, 342 (1966)

7. *Boson-Broadened Photonuclear Reactions in Light Nuclei,*
   C B Duke, F B Malik and F W K Firk,
   Phys. Rev. **157**, 879 (1967)

8. *Energy Dependence of Isospin Mixing in the Giant Dipole States of C^{12} and O^{16}*
   C-P Wu, F W K Firk and T W Phillips,
   Phys. Rev. Letts. **20**, 1182 (1968).

9. *The Differential Cross Sections of Photoprotons from ^{16}O in the Region of the Giant Dipole Resonances*
   J E E Baglin and M N Thompson,
   Nucl. Phys. **A158**, 73 (1969)

10. *Total Neutron Cross Section Measurements*
    F W K Firk and E Melkonian, in
    Experimental Neutron Resonance Spectroscopy,
    Ed. J A Harvey, Academic Press, New York
    101 (1970)

11. *Low-Energy Photonuclear Reactions*
    F W K Firk in Annual Review of Nuclear
    Science, Vol. **20** (Annual Reviews Inc.
    Palo Alto), 39 (1970)

12. *Measurement of Total Photonuclear Cross
    Sections Using a Novel Method of Photon
    Spectroscopy*
    C-P Wu, B L Berman and F W K Firk
    Nucl. Instr. and Meths. **79**, 346 (1970)

13. *Differential Cross Section of Photoprotons from
    Deuterium*, B Weissman and H L Schultz,
    Nucl. Phys. **A174**, 129 (1971)

14. *Polarization Studies Using the Neutron Spin Precession Method with a Continuous Energy Spectrum of Neutrons*

R Nath, F W K Firk, R J Holt and H L Schultz
Nucl. Instr. and Meths. **98**, 385 (1972)

15. *Polarization of Neutrons in n-^{12}C Scattering: A Standard for Polarization Studies in the MeV Region*

R J Holt, F W K Firk, R Nath and H L Schultz,
Phys. Rev. Letts. **28**, 114 (1972)

16. *The Photonuclear Future*

P Axel, W Bertozzi, F W K Firk, M Goldhaber,
W Greiner and E Teller,

Proc. Inter. Conf. on Photonuclear Reactions
and Applications, Vol. 1,

Ed. B L Berman (USAEC Office of Info. Servs.,
Oak Ridge) 155 (1973)

17. *R-Matrix and Phase Shift Analyses of Neutron Polarization Measurements from n-^{16}O Scattering*

G T Hickey, F W K Firk, R J Holt and R Nath
Nucl. Phys. **A225**, 470 (1974)

18. *The Polarization of MeV Neutrons Elastically Scattered from ^4He*

   J E Bond and F W K Firk,

   Nucl. Phys. **A258**, 189 (1976)(1975)

19. *Neutron Polarization*

   F W K Firk, Proc. Inter. Conf. on the

   Interaction of Neutrons with Nuclei, Vol. 1

   Lowell, Ed. E Sheldon (USERDA Tech. Info.

   Center, Oak Ridge) 389 (1976)

20. *Neutron Time-of-Flight Spectrometers*

   F W K Firk, Nucl. Instr. and Methods,

   162, 237 (1979)

21. *N-P Capture and the Photodisintegration of the Deuteron*, F W K Firk in Neutron Capture

   Gamma Ray Spectroscopy, Eds. R E Chrien and

   W R Kane (Plenum Press, New York) 701 (1979)

22. *The Polarization and Differential Cross Section of Fast Neutrons Scattered from ^6Li and an R-Function Analysis of the Interaction below 4 MeV*

   Y-H Chiu and F W K Firk,

   Nucl. Phys. **A364**, 43 (1981)

23. *Polarization Effects in the Small-Angle Scattering of Fast Neutrons from Bismuth*

M Ahmed and F W K Firk

Proc. International. Symp. on Polarization Phenomena in Nuclear Physics, Santa Fe, Eds. G G Ohlsen et al (American Inst. of Physics, New   York) 389 (1981)

23. *Transverse and Longitudinal Isoscalar and Isovector modes in Nuclear Giant Resonances*

A Vazquez, Phys. Rev. Lett. **50**, 1756 (1983)

24. *A study of M1, E1 and E2 strengths in ^{16}O below 25 MeV*

F W K Firk, J. Phys. G: Nucl. Part. Phys. **17**, 1739 (1991)

25. *Differential polarization of photoneutrons from ^{16}O*

F W K Firk, J. Phys. G: Nucl. Part. Phys. **21**, 1341 (1995)

8. Summary

An overview of the pioneering research carried out at the Electron Accelerator Laboratory at Yale University over a period of four decades has been given. The fields covered include neutron and photon interactions with nuclei. 44 Ph. D's were awarded in these two fields. Studies of elastic and inelastic scattering of electrons with nuclei were not reviewed. Research in these fields was, and is, extensive and a separate book would be required to do justice to the work. 8 Ph. D's were awarded in electron scattering in an 8-year period.

The government funding of the research at the Laboratory suddenly stopped in late 1983. The reason clearly had to do with the cost (> 1 million dollars) needed to construct an improved Tandem Van de Graaff accelerator in the nearby Wright Nuclear Structure Laboratory. The government could not support both projects. The electron accelerator was dismantled over a period of time, and many of its parts were shipped to a sister accelerator at RPI, where they were put to good use. It was the end of an important era in the history of Physics at Yale.

www.ingramcontent.com/pod-product-compliance
Lightning Source LLC
Chambersburg PA
CBHW071214220526

45468CB00002B/600